DOWN IN THE OCEAN

TIDE POOL TREASURES

BY MELISSA GISH

CREATIVE EDUCATION • CREATIVE PAPERBACKS

Published by Creative Education and Creative Paperbacks
P.O. Box 227, Mankato, Minnesota 56002
Creative Education and Creative Paperbacks are imprints of
The Creative Company
www.thecreativecompany.us

Design, production, and illustrations by Chelsey Luther
Art direction by Rita Marshall
Printed in China

Photographs by Alamy (David Fleetham, Steven J. Kazlowski, National Geographic Creative, Sanamyan), All-free-download.com, Flickr (Jerry Kirkhart), Getty Images (Gerald and Buff Corsi/Visuals Unlimited, Inc., David Hall/Science Source, robertharding, ALEXANDER SEMENOV/SCIENCE PHOTO LIBRARY, Dean vant Schip/All Canada Photos, Art Wolfe/The Image Bank), iStockphoto (andipantz, AvatarKnowmad, BrendanHunter, connerscott1, GeorgeBurba, leezsnow, lovleah, milynmiles, NNehring, trekandshoot, Velvetfish, Zampelli, Michael Zeigler), Minden Pictures (Ross Hoddinott/NPL, Constantinos Petrinos/NPL, Nick Upton/NPL), Shutterstock (a2_photography, Chase Clausen, Zacarias Pereira de Mata, ItsAngela, Matt Knoth)

Copyright © 2019 Creative Education, Creative Paperbacks
International copyright reserved in all countries. No part of this book may be reproduced in any form without written permission from the publisher.

Library of Congress Cataloging-in-Publication Data
Names: Gish, Melissa, author.
Title: Tide pool treasures / Melissa Gish.
Series: Down in the ocean.
Includes bibliographical references and index.
Summary: Explore the regions of the world's oceans known for their intertidal zones and learn about the life forms that dwell there. First-person accounts from scientists answer important questions about tide pool ecosystems.
Identifiers: LCCN 2017028054 / ISBN 978-1-64026-000-9 (hardcover) / ISBN 978-1-62832-555-3 (pbk) / ISBN 978-1-64000-029-2 (eBook)

Subjects: LCSH: 1. Tide pool ecology—Juvenile literature. 2. Tide pool animals—Juvenile literature.
Classification: LCC QH541.5.S35 G57 2018 / DDC 577.69/9—dc23

CCSS: RI.4.1, 2, 7; RI.5.1, 2, 3, 8; RST.6-8.1, 2, 5, 6, 8

First Edition HC 9 8 7 6 5 4 3 2 1
First Edition PBK 9 8 7 6 5 4 3 2 1

TABLE OF CONTENTS

Welcome to the Tide Pool	4
Infinite Wonders	7
Eat or Be Eaten	13
Special Relationships	19
Family Life	25
Ocean Mysteries	31
True-Life Tide Pool Adventure	36
Under Pressure	41
Glossary	46
Selected Bibliography	47
Index	48

WELCOME TO THE TIDE POOL

The ocean level rises and falls along coastlines every day. These changes are called tides. The habitat along coastlines is called the intertidal zone. It has four parts. The low-tide zone is underwater most of the time. It can be seen only when the tide is lowest. Animals that live here include sponges and sea hares.

Rocky coasts hold water in tide pools in the mid-tide zone. Sea stars, sea anemones, crabs, and fish live in these pools. The high-tide zone is not often underwater. Barnacles and other hard-shelled animals can survive here. The splash zone is where the tide reaches only during storms. Here, crabs and shorebirds may find dead plants and animals to eat. Intertidal zones are amazing ocean **ecosystems**.

INFINITE WONDERS

Tide pools form when seawater is trapped on the shore by rocks. Some tide pools are shallow. Others can be quite deep. For part of each day and night, tide pool inhabitants are exposed to wind, waves, predators, and changes in water temperature. The organisms that live in tide pools must be able to cope with these changes.

Some animals, such as oysters and barnacles, are sessile. This means they are fixed to a surface. They cannot move. Other animals swim about or climb over rocks. Sea stars, urchins, and crustaceans hide under rock ledges. Colorful nudibranchs (*NOO-dih-branks*) creep slowly. Octopuses dart with lightning speed. Life in a tide pool is hazardous and unpredictable. But it is also rewarding. Each time the tide returns, it carries precious food to the animals in the intertidal zone.

Varied and vibrant

Nudibranchs are also known as sea slugs. They have tubular, rounded, or flattened growths on their bodies called cerata. These collect and help digest food. They also function as gills. Some nudibranch species have only a few cerata. Others have many.

Building sandcastles

Hundreds of sandcastle worms gather together to form colonies. When covered by water, their feathery tentacles capture food and grains of sand. They glue the sand together to make protective tubes. At low tide, they seal themselves inside their tubes.

ASK A  SCIENTIST

Do plants live in tide pools?

Yes, they do. One special kind of plant, called surfgrass, can often be found in tide pools. It forms long, bright green leaves. During low tide, when the water flows back to the ocean, tide pools heat up and become stressful places. Surfgrass helps the other organisms in tide pools by providing a dark, moist environment for them.

— Dr. Kevin A. Hovel, Marine Biologist, San Diego State University

Hilton's aeolid nudibranch

sandcastle worm colony

Licking up lunch

Abalone are snails. They cling to rocks with a thick muscle called a foot. When seaweed drifts by, they trap it. Then they use their tongue-like radula to scrape the seaweed into bits. The bits are then swallowed. Most mollusk species have a radula.

Sunflower of the sea

The sunflower star is the largest sea star in the world. Young sunflower stars have five arms, called rays. They grow more rays as they age, ending up with 24. They can move almost 10 feet (3 m) per minute. This is super-speedy for a sea star.

ASK A  SCIENTIST

How do tide pool species cope with low tide?

The key is that they need to stay moist and not get too hot (or too cold in the temperate zone). They stay wet by closing up (barnacles, anemones) or staying under rocks and in crevices that don't dry out. The higher up in the intertidal they are, the more time they are exposed. Barnacles are one of the best at surviving exposure.

— Dr. Philip Hastings, Marine Biologist, University of California, San Diego

2

EAT OR BE EATEN

Algae and seaweed are important in the tide pool food web. They make their own food from the sun's energy. Many herbivores, such as limpets and sea urchins, feed on algae and seaweed. Other tide pool animals are predators. They hunt their neighbors. In a tide pool, options are limited. There may be nowhere to flee when attacked by a predator. To survive, many creatures have adapted methods of protection. But for every animal's means of survival, another has a means of attack. Tide pools are harsh living spaces.

I smell danger!

Tidepool sculpins can smell their main predator, kelp greenlings. They can also smell injured sculpins. This lets them know that danger is nearby. They usually settle into the sediment and are well camouflaged. But these fish can also breathe air to survive during low tide.

Clam soup

The moon snail grabs clams and drags them into the sand. Using its radula, the snail scrapes a small hole in the clamshell. Then it squirts acid into the clam. This turns the clam to soup. The snail then slurps up its prey.

ASK A  SCIENTIST

Are there fierce predators in a tide pool?

In the tide pool, sea stars are among the fiercest predators. They prefer to prey on **bivalves**. They use their tube feet to pry open the bivalve shell. Then they can slip their stomach out of their mouth and in between the small gape of the shell. As the bivalve gets slowly digested, it gets weaker. The shell gape becomes wider, allowing the sea star to fully consume its prey.

— Dr. Nancy Smith, Marine Scientist, Eckerd College

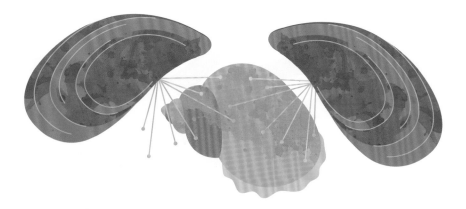

Superhero mussels

Blue mussels are often targeted by hungry dog whelks. When these snails attack, the mussel may defend itself like Spider-man! It shoots the snail with byssal threads, which are produced by the muscular foot. These threads hold the snail in place like spider silk.

ASK A 🐙 SCIENTIST

What do you think is the most interesting living thing in a tide pool?

The ragworm. It is a worm as long as your hand with stubby legs that run up and down its body. It looks like a torn piece of fabric. It has big jaws to catch prey, and when it wriggles, it makes beautiful colors. Some people use it for fish bait.

— Dr. David Samuel Johnson, Marine Ecologist, Virginia Institute of Marine Science

3

SPECIAL RELATIONSHIPS

Tide pools are small, dangerous worlds. Algae-covered rocks provide minimal protection. During low tide, many hiding places are exposed. Most animals do their best to avoid one another. But some have formed special relationships. Partners may share space or skills with each other. Such an arrangement gives something good to both partners. Other times, one animal may simply take advantage of another. No one is harmed, and it turns out to be useful. Many tide pool creatures rely on others.

SYMBIOTIC RELATIONSHIP EXAMPLES

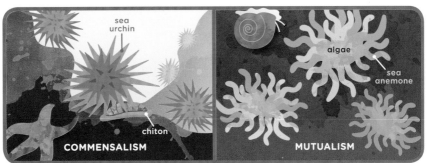

Munch on me

Coralline algae can be smothered by other algae growing on top of them. So coralline algae may shed their top layer to push off the intruders. They also welcome herbivores that eat the top layer of algae. This frees the coralline algae to grow again.

Spiny bodyguards

When chitons (*KY-tunz*) are pulled from a rock, they curl up in a ball. They resemble the insects commonly called roly-poly bugs. Large sea stars hunt chitons. Some chitons live among groups of purple sea urchins. The urchins help protect the chitons from sea stars.

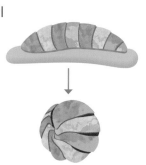

ASK A 🐙 SCIENTIST

Are algae the same as plants?

Even though plants such as surfgrass can be found in tide pools, most of the things seen in tide pools that look like plants actually are algae. Algae do many things that plants do, but they are a completely different type of organism. They can be all sorts of colors, from red to brown to green and many others. Algae are important for providing a home and food for many animals.

— Dr. Kevin A. Hovel, Marine Biologist, San Diego State University

breadcrumb sponge

green anemone

Is that moldy bread?

The breadcrumb sponge is yellow, but it often looks green. That is because zoochlorellae (*ZOH-uh-klor-EL-uh*) algae live on the sponge. The algae provide nutrients to the sponge, helping it grow. And the sponge gives the algae a safe place to live.

ASK A SCIENTIST

What makes the sea anemone special?

It looks like a plant, but it is not. It also looks like a sun, giving rocks pretty colors. Sea anemones eat many different things, including **detritus**, which is important for recycling. They sting when touched (except for some fishes). This is to catch small animals to eat. Many sea anemones are also fluorescent at night when viewed with a black light. They are magical to look at!

— Dr. Dimitri Deheyn, Marine Biologist, Scripps Institution of Oceanography

FAMILY LIFE

Reproduction is easy for some tide pool animals. For others, it can be a challenge. Sponges do not need mates. They simply split and create copies of themselves. Some animals, such as sea slugs, have both male and female reproductive organs. They can mate with either gender. Some creatures glue clusters of eggs to rocks. Others release eggs that drift away at high tide. When the young animals, called larvae, hatch, most become food for larger organisms. The ones that survive can populate more tide pools.

TIDE POOL REPRODUCTION

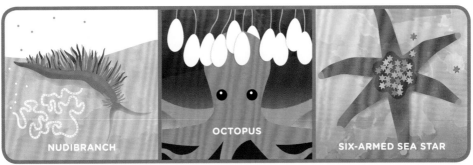

NUDIBRANCH OCTOPUS SIX-ARMED SEA STAR

Blind dates

Barnacles are crustaceans living inside cone-shaped shells. The shells are glued to surfaces, so barnacles cannot move around. To find mates, male barnacles extend their reproductive organs out of their shells. They blindly poke around neighboring barnacles until they find a female.

Family meals

Frilled dog whelks are sea snails. Females attach their eggs to rocks. Sometimes, they eat their own eggs! When the baby whelks hatch, the first ones to emerge may eat the eggs of their unhatched siblings.

ASK A SCIENTIST

What is your favorite tide pool creature?

The six-armed sea star! This tiny sea star has eyespots at the tip of each arm. These primitive eyes help locate prey. During winter and spring, the females produce large, sticky eggs that they assemble into a mound attached to the rock. They **brood** the eggs for several months, keep them clean, and protect their offspring until they are large enough to crawl away as tiny sea stars.

— Dr. Steven Rumrill, Biologist, Oregon Department of Fish and Wildlife

ASK A 🐙 SCIENTIST

What makes chitons special?

Many chitons exhibit homing behavior in which they return to the exact same spot during the day after roaming around at night to feed. Chitons have teeth that they use to scrape algae on the rocks, so it's easy to trace their nighttime movements. It's amazing that they can always find their way back to their "home." I've found that some chitons love the same spot so much that over time, they have carved out a small depression on the rock that perfectly fits their oval body.

— Dr. Nancy Smith, Marine Scientist, Eckerd College

Rock-climbing fish

The common blenny, or shanny, hides under rocks or seaweed at low tide. It can breathe out of the water during this time. The blenny uses its side fins to climb over rocks. Females lay eggs under abandoned shells. Males guard the eggs until they hatch.

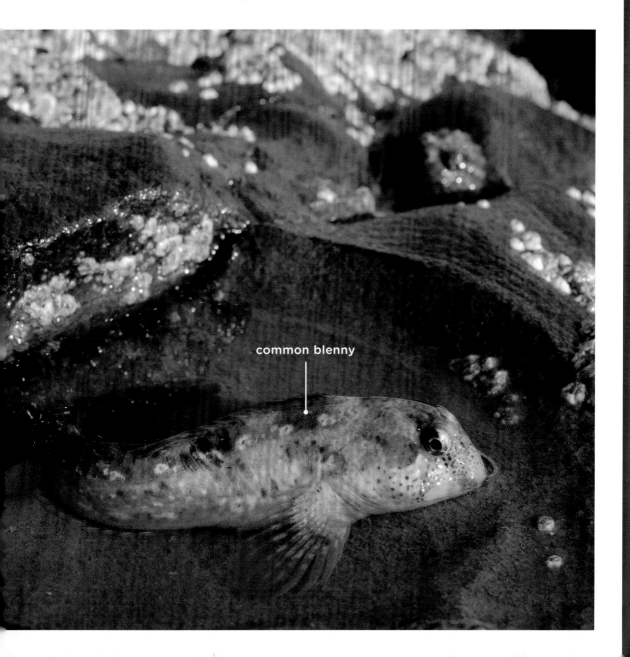

common blenny

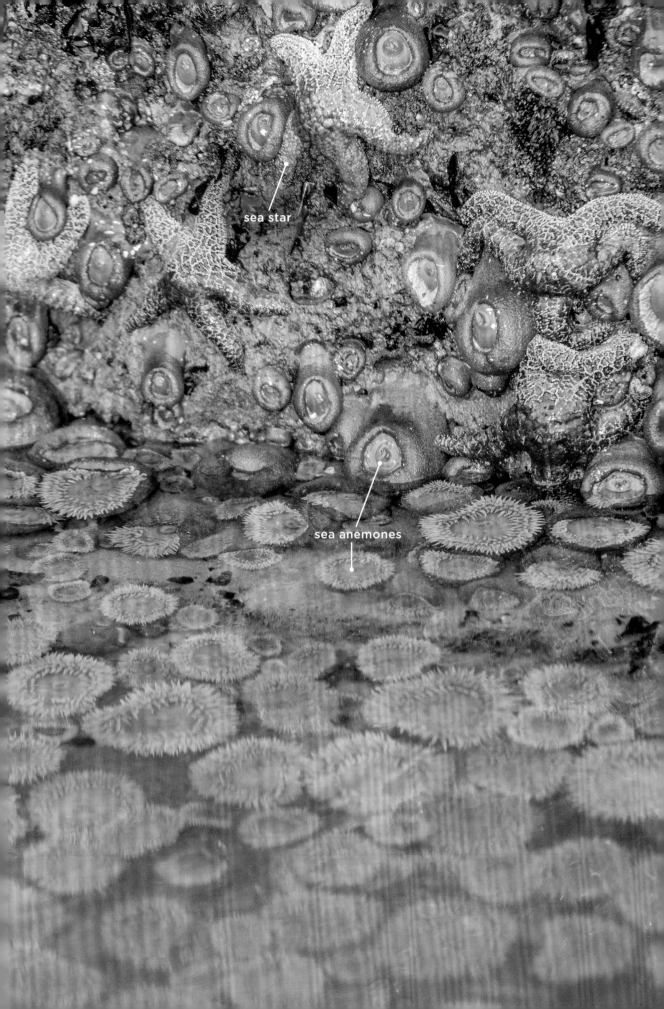

5

OCEAN MYSTERIES

California is famous for its tide pools. But tide pools exist all over the world. They form around the edges of warm, tropical islands such as Mexico's Puerto Peñasco and the Solomon Islands in the South Pacific. They are found in cool climates, such as southern England's Wembury Bay and Vancouver Island's Botanical Beach. Tide pools even exist in the Arctic—though no plants and few animals are found in them.

Most tide pools are home to some of the ocean's most unusual creatures. Chitons and barnacles can withstand powerful ocean waves. Anemones blossom like flowers at high tide and shrivel like rocks at low tide. Opalescent and clown nudibranchs, sea lemons, and Hudson's dorids look like aliens from another world. Algae and encrusting sponges appear to paint rocks with brilliant colors. Every visit to a tide pool can reveal new secrets.

Star systems

Golden star tunicates are about 0.1 inch (2.5 mm) long. Many of these animals join together in a group called a system. Most systems are about two inches (5.1 cm) long. Some systems form double rows. Others are star-shaped. Tunicates are also called sea squirts.

Stand-up seaweed

Most seaweed is floppy outside the water. But the palm seaweed has a stiff stalk that holds up its leaflike fronds. It looks like a small palm tree. Like algae and other seaweeds, the palm seaweed makes its own food from the sun's energy.

ASK A  SCIENTIST

Can sea stars really regrow limbs?

Sea stars can regrow arms that are lost to predators or other kinds of damage. In fact, some sea stars reproduce by splitting in half and letting each half regrow the missing parts. In most cases, a sea star must have at least part of the body to replace the arms. A few, however, are able to grow an entirely new body and four arms from a single arm.

— Dr. Richard L. Turner, Marine Biologist, Florida Institute of Technology

golden star tunicate system

palm seaweed

hermit crab

California two-spot octopus

Cleanup crew

Hermit crabs live in empty snail shells to protect their soft abdomens. As they grow, they need to find bigger shells. Their two front claws are used for crushing and eating food. They mostly eat dead things. This helps keep their tide pool habitats clean.

Ambush hunter

Shallow-water octopuses hide under rock ledges in tide pools. They may dash out of the water to grab passing crabs. They must drag their prey underwater to feed. An octopus has a sharp beak shaped like that of a parrot. It can crush the crab's shell.

ASK A  SCIENTIST

Do you have a favorite marine animal?

Yes! Nudibranchs, also known as sea slugs. They are cousins of garden snails, but they have lost their shells, so their gills are exposed, or naked. The word *nudibranch* means "naked gills" in Latin. They are incredibly colorful, and every time I see them, I am amazed about the combination of the colors that each species has developed along their **evolution**.

— Dr. Octavio Aburto Oropeza, Marine Biologist, Scripps Institution of Oceanography

bat star

TRUE-LIFE TIDE POOL ADVENTURE

A DAY AT THE SAN DIEGO TIDE POOLS

Allison Randolph grew up in South Florida. Almost from birth, she knew she was meant for a life connected to the sea. She learned to swim, snorkel, and surf. She often sailed with her parents between Florida and the Bahamas. Later, she moved to California. There she discovered that one of her favorite ways to spend a day was visiting the tide pools in San Diego. She always made sure to arrive about one hour before low tide. That way, she would have plenty of time to explore before the tide returned.

One day, Allison glanced around and saw mostly rocks and seaweed. But when she bent down and took a closer look, a fascinating world revealed itself to her. She gently pushed a bit of seaweed aside. There was a beautiful, green starburst anemone eating a shore crab! She slowly lifted a rock and looked underneath. There was an orange-pink bat star! She carefully put the rock back exactly as she'd found it. Next, she gazed in a shallow pool of water surrounded by rocks. There was her favorite tide pool animal: a Spanish shawl nudibranch. It was only about 1.5 inches (3.8 cm) long, but it was unmistakable. Bright orange feather-like appendages fluttered along a pink body. Allison looked into another pool. There was a two-spot octopus. Its head was about the size of a grapefruit, and its arms swirled as it chased a fish around the pool.

Allison continued to explore. She carefully moved bits of seaweed and lifted small rocks (always being sure to replace them). She found brittle stars. Some were a half-inch (1.3 cm) across. Others were as many as six inches (15.2 cm) wide. She watched a *Navanax* creep along the end of a rock. This tiny sea slug was brown with teal spots. Next, she gently touched a sea hare, one of the squishiest animals she had ever felt. The tide was coming back in. Allison retreated from the tide pools with a big smile on her face. "My day was filled with color and excitement!" said Allison.

Spanish shawl nudibranch

6

UNDER PRESSURE

Human activities, such as the burning of fossil fuels, are putting carbon dioxide gas into the oceans. Mixed with seawater, the gas turns into carbonic acid. This is the substance that makes soda fizzy. Scientists call this condition "ocean acidification." It has upset the chemical balance of ocean water. Scientists are rushing to study it.

Some tide pool animals affected by acidification are mussels, abalone, and sea urchins. Their shells are now thinner. Their bodies are weaker, and they struggle to compete for food. Acidification also affects animals' nervous systems. Some tide pool snails cannot sense sea stars. They do not try to escape when attacked. We must study the changes in tide pools so that we can protect these unique habitats. Tide pool ecosystems have important ties to all organisms that live down in the ocean.

ASK A 🐙 SCIENTIST

Do hurricanes hurt tide pools?
How soon can tide pools recover from hurricanes?

Hurricanes hurt tide pools. [As ocean temperatures rise, stronger hurricanes and typhoons batter the shores.] In fact, they are some of the more severe disturbances to tide pools in areas where the hurricanes come ashore. The huge waves can be very destructive to the animals and plants that live in them. But animals and plants in the sea recover pretty quickly, and soon they will rebound to their original condition. This may take weeks to months for many seaweeds and crawling animals, or maybe a couple years for long-lived animals like corals—but they will recover.

— Dr. Matthew Edwards, Marine Biologist, San Diego State University

Buried by a tsunami

A 2011 tsunami changed Japan's Mangoku-ura Lagoon. Tide pools filled with mud. Plants took over. Mussels and clams were smothered. Today, scientists are studying how the tide pools are still struggling to recover.

Abalone Cove

Protecting California's tide pools

Since 2010, California has made dozens of tide pool habitats into marine protected areas. Some of these include Del Mar Landing, Abalone Cove, and Duxbury Reef. No one is allowed to disturb creatures in these areas.

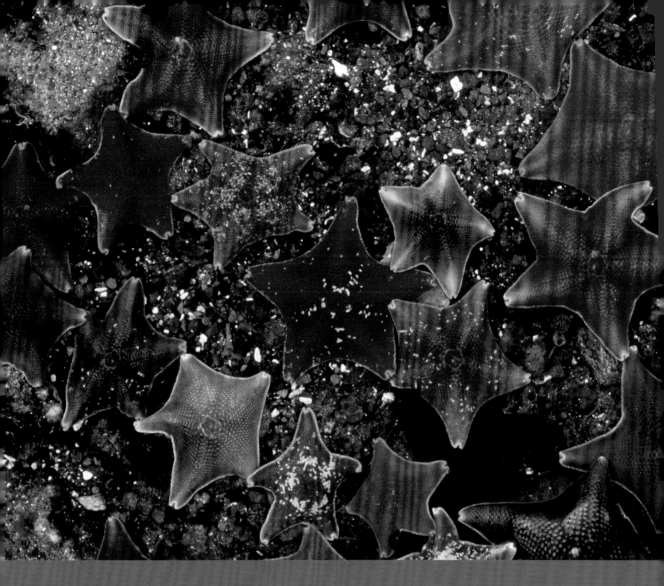

ASK A 🐙 SCIENTIST

What is something kids can do to help protect tide pools?

Introduce your friends to tide pools. Take photos of neat organisms. Revisit your tide pool on a regular basis. Pick something to count, count it, and keep track over time, seasonally and year to year. Citizen scientists are in a position to collect data that may be early signs of **deleterious** change. Even better, pick two things to count (A and B). Track their relative counts (A/B) over time. For example, count sea anemones and sea stars.

— Dr. Terry Gaasterland, Marine Biologist, Scripps Institution of Oceanography

GLOSSARY

adapted
changed to improve its chances of survival in a changed environment

bivalves
aquatic soft-bodied animals that grow inside a two-piece hinged shell

brood
to care for and protect eggs

camouflaged
hidden by blending into the shape and color of the environment

chemical
a substance created or changed by a natural scientific process

climates
the weather conditions that remain mostly the same over a long period of time

crustaceans
animals with no backbone that have a shell covering a soft body

deleterious
causing harm or damage

detritus
waste matter, especially from decomposing organisms

ecosystems
communities of organisms that live together in balance

evolution
the process of gradual development into a new form

food web
a system in nature in which living things depend on one another for food

gills
body parts that extract oxygen from water

mollusk
a member of a large group of spineless animals that includes snails, slugs, mussels, clams, and octopuses

species
a group of living beings with shared characteristics and the ability to reproduce with one another

tentacles
slender, flexible limbs in an animal, used for grasping, moving about, or feeling

SELECTED BIBLIOGRAPHY

Eunice, Kasey. *Life in a Tide Pool* [series]. Aliso Viejo, Calif.: Splashzone Productions, 2015.

ExplorOcean. "Tide Pools." *YouTube*. https://youtu.be/ZkULAD8gJT0.

Fylling, Marni. *Fylling's Illustrated Guide to Pacific Coast Tide Pools*. Berkeley, Calif.: Heyday Books, 2015.

National Geographic Society. "Tide." https://www.nationalgeographic.org/encyclopedia/tide/.

Oregon State Parks. "Video Gallery." *Oregon Tide Pools*. http://oregontidepools.org/videogallery.

Tway, Linda E. *Tidepools*: *Southern California: A Guide to 92 Locations from Point Conception to Mexico*. 2nd ed. Birmingham, Ala.: Wilderness Press, 2011.

Note: Every effort has been made to ensure that any websites listed above were active at the time of publication. However, because of the nature of the Internet, it is impossible to guarantee that these sites will remain active indefinitely or that their contents will not be altered.

INDEX

algae 13, 19, 20, 21, 23, 28, 31, 32
barnacles 4, 7, 11, 26, 31
chitons 20, 28, 31
clams 15, 43
common blennies (shannies) 29
corals 42
crabs 4, 35, 37
defenses 13, 15, 16, 20, 23, 35
    camouflage 15
    stinging 23
diets 4, 7, 8, 11, 13, 15, 16, 20, 23, 25, 26, 28, 32, 35, 37, 41
    detritus 23
    herbivorous 13, 20
    and photosynthesis 13, 32
    prey 15, 16, 20, 23, 25, 26, 35, 37, 41
    scavenging 4, 23, 35
golden star tunicates (sea squirts) 32
hermit crabs 35
high tide 4, 25, 31
kelp greenlings 15
low tide 4, 8, 11, 15, 19, 29, 31, 37
marine protected areas 44
mussels 41, 43
octopuses 7, 35, 37
oysters 7
predators 7, 13, 15, 16, 20, 23, 25, 32, 35, 37, 41

ragworms 16
Randolph, Allison 36–37
relationships 8, 13, 19, 20, 23, 35
    commensalism 20
    mutualism 19, 20, 23, 35
reproduction 25, 26, 28, 32
    eggs 25, 26, 29
    and offspring 25, 26
sandcastle worm colonies 8
scientists 8, 11, 15, 16, 21, 23, 26, 28, 32, 35, 41, 42, 43, 45
    David Samuel Johnson 16
    Dimitri Deheyn 23
    Kevin A. Hovel 8, 21
    Matthew Edwards 42
    Nancy Smith 15, 28
    Octavio Aburto Oropeza 35
    Philip Hastings 11
    Richard L. Turner 32
    Steven Rumrill 26
    Terry Gaasterland 45
sea anemones 4, 11, 23, 31, 37, 45
sea slugs 4, 7, 8, 25, 31, 35, 37
sea snails 11, 13, 15, 16, 26, 35, 41
    abalone 11, 41
    dog whelks 16, 26
    limpets 13
    moon snails 15
sea stars 4, 7, 11, 15, 20, 26, 32, 37, 41, 45

    bat 37
    brittle 37
    six-armed 26
    sunflower 11
sea urchins 7, 13, 20, 41
seaweed 11, 13, 29, 32, 37, 42
sponges 4, 23, 25, 31
surfgrass 8, 21
threats 41, 42, 43, 45
    acidification 41
    climate change 42, 43
tide pool locations 4, 11, 31, 36, 37, 43, 44
    Arctic 31
    California 31, 36, 37, 44
    Canada 31
    England 31
    Japan 43
    Mexico 31
    Solomon Islands 31
tide pool terrain 4, 7, 11, 19, 20, 23, 25, 26, 28, 29, 31, 35, 37
    and sessile inhabitants 7, 19, 23, 26, 31
    zones 4, 7, 11
tidepool sculpins 15